THE COLOUR
BUBBLECARS
& MICROCARS
FAMILY ALBUM

This book is dedicated to John & Dulcie Collins.

Other books of interest to enthusiasts available from Veloce -

Alfa Romeo Owner's Bible by Pat Braden
Alfa Romeo Tipo 6C, 1500, 1750 & 1900 by Angela Cherrett
Alfa Romeo Modello 8C 2300 by Angela Cherrett
Alfa Romeo Giulia Coupé GT & GTA by John Tipler
Biggles! by Peter Berresford Ellis & Jennifer Schofield
British Car Factories from 1896 - A Complete Survey by Paul Collins & Michael Stratton
Bugatti 57 - The Last French Bugatti by Barrie Price
Car Bodywork & Interior: Care & Repair by David Pollard
Car Security Manual by David Pollard
Citroën 2CV Family Album by Andrea & David Sparrow
Citroën DS Family Album by Andrea & David Sparrow
Cobra: The Real Thing! by Trevor Legate
Completely Morgan: Four-Wheelers from 1936 to 1968 by Ken Hill
Completely Morgan: Four-Wheelers from 1968 by Ken Hill
Daimler SP250 'Dart' by Brian Long
Fiat & Abarth 124 Spider & Coupé by John Tipler
Fiat & Abarth 500 & 600 by Malcolm Bobbitt
Lola T70 by John Starkey
Mazda MX5/Miata Enthusiast's Workshop Manual by Rod Grainger & Pete Shoemark
Mini Cooper: The Real Thing! by John Tipler
Nuvolari: When Nuvolari Raced ... by Valerio Moretti
Pass the MoT by David Pollard
The Prince & I - My Life with the Motor Racing Prince of Siam (biography of racing driver 'B.Bira') by Princess Ceril Birabongse
Standard & Standard-Triumph - The Illustrated History by Brian Long
Total Tuning for the Classic MG Midget/A-H Sprite by Daniel Stapleton

First published in 1994 by Veloce Publishing Plc., Godmanstone, Dorset DT2 7AE, England. Fax 0300 341065

SEMINOLE LIBRARY ISBN 1 874105 29 4

© Andrea Sparrow, David Sparrow and Veloce Publishing Plc 1994

All rights reserved. With the exception of quoting brief passages for the purpose of review, no part of this publication may be recorded, reproduced or transmitted by any means, including photocopying, without the written permission of Veloce Publishing Plc.

Throughout this book logos, model names and designations, etc., may have been used for the purposes of identification, illustration and decoration. Such names are the property of the trademark holder as this is not an official publication.

Readers with ideas for automotive books, or books on other transport or related hobby subjects, are invited to write to the editorial director of Veloce Publishing at the above address.

British Library Cataloguing in Publication Data -
A catalogue record for this book is available from the British Library.

Typesetting (Avant Garde), design and page make-up all by Veloce on Apple Mac.

Printed in Hong Kong

THE COLOUR BUBBLECARS & MICROCARS FAMILY ALBUM

ANDREA & DAVID SPARROW

VELOCE PUBLISHING PLC
PUBLISHERS OF FINE AUTOMOTIVE BOOKS

THANKS

Our grateful thanks are due to the following: Jade Bond, Sara Bloyce, Jean Hammond, Mike Webster, Sjoerd Ter Burg, Dick Boerstoel, The Auto Trom Museum in Rosmalen, Patrick Pellen, David Hamilton, Colin Archer, Ron Crawley, Lawrence House, Malcolm Thomas, Steve Dorman, Steve Hurn, Jose Zijerveld, Winfred Jones, Bob Parry, the members of the German Micro Car Club, the members of the Dutch Bubble Car Club "DWAC". A special thank you to the Brooklands Motor Museum

A very special thank you to Henk Kleinendorst for whom no request was too much trouble and whose good-humoured enthusiasm has been invaluable.

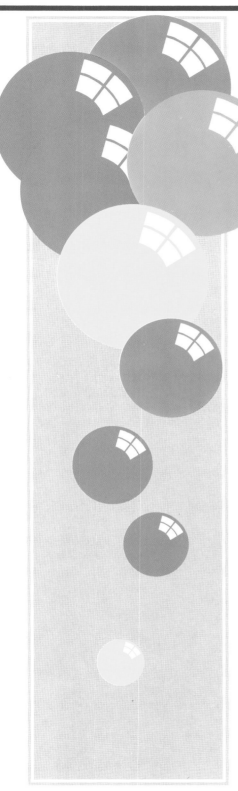

CONTENT

	INTRODUCTION	6
1	HEINKEL & TROJAN	7
2	JANUS	23
3	ISETTA	29
4	LADYBIRD	47
5	MESSERSCHMITT	55
6	PEEL TRIDENT	76
7	SHELTER	82
8	VELOREX	87
	POSTSCRIPT	96

INTRODUCTION

In their reviews and road tests, the motoring journalists of the fifties had a clear idea of what they wanted from small car motoring -

"With fuel scarce and expensive, a well-contrived small car must undoubtedly benefit given three conditions: that it shall fill the position of a larger car in reasonable comfort and performance, and that it shall still be economical when driven to keep up with normal traffic."
(Motor - 1957)

"Time and again it has been stressed that these miniatures should not be just scaled-down automobiles. To get the best results, quite novel approaches have to be found, especially in respect of weight disposition and suspension." (Autocar 1956)

In Britain, three was the magic number for wheels. Three-wheeled cars paid road tax at the same low rate as motorcycles and sidecars and also attracted a lower rate of purchase tax. So one finds original British-designed cars starting life with three wheels and imported designs, which originally started with four wheels, being modified to suit, as with the Isetta and Heinkel/Trojan.

The bubblecar and microcar international family is huge: this book does not profess to cover even the tip of the iceberg. Neither does it aim to be an encyclopaedia of makes or models. Rather it is the first volume of a colourful photographic celebration of the world's fascinating and, often, visually eccentric microcars. Miniature car motoring is not for the shy and retiring: take a trip in a bubblecar and watch people turn, stare and, sometimes, laugh.

The names of some of the cars featured in this book are well-known like Messerschmitt (of which only the three-wheelers are included here), Isetta (the BMW line only), and Heinkel/Trojan. Others are less well known: the Zundapp Janus from Germany, Velorex from Czechoslovakia, Shelter from Holland and Peel Trident from the Isle of Man. The British BSA Ladybird is unique.

The cars in this book were the solution to the quest for inexpensive, liberating motoring in the middle of the twentieth century. With this one major point in common, they are perhaps less like a family, more like club - a group of charismatic individuals linked together by a common purpose.

Andrea Sparrow

HEINKEL & TROJAN

Heinkel of Stuttgart was, like so many others in the bubblecar field, primarily an aircraft building company, though it had been producing motor scooters since 1953. In 1955 the company introduced a three-wheeled car, called the *Kabinen*, (Cabin Cruiser) which was similar in appearance to the Isetta and was powered by a 174cc single-cylinder four-stroke engine. However, unlike the Isetta, the Kabinen had no chassis. Whilst the new car's rack and pinion steering was highly praised, as was its ride, Ernst Heinkel did have a problem with rainwater: it tended to get inside the car and slosh around on the floor. He asked his engineers why Heinkel aeroplanes did not suffer from the same problem. The engineers explained that they had found a simple method to keep the 'planes water-free - they drilled holes in the floors to let it out!

A 204cc engine was introduced in 1956 and, at the same time, twin rear-wheels were fitted instead of one. The interior design of the Heinkel offered room for two adults at the front and two children in the rear to travel in comfort; smallish children, because the rear had insufficient headroom for anyone approaching normal adult size. One criticism of the seating was that it was rather sternly upright, and gave a taller driver backache after a while. The front opening door hinged at the side, although the steering column was fixed rather than lifting with the door. There was little possibility of parking head-on to the kerb, as the car was longer than the width of a more conventionally parked vehicle. The car was felt to be comfortably stable under most conditions, although downhill on a bad surface was somewhat hairy!

Travelling in the car today, the overriding impression is that if you're not deafened by the noise, then, at very least, your tooth fillings will be vibrated out. Engine noise is amplified by the 'dome' and given a special harshness by the nature of the large Plexiglass area. With the roof open, the noise level is at least cut down (or perhaps exchanged for wind noise?).

In its time the car was considered the prettiest of the

Below: The original badge of Heinkel in Germany was replaced when production began in Northern Ireland by a similar logo with an added "I". When Trojan of Croydon, England, started production, the logo changed completely.

Twin rear wheels were fitted to German-produced cars from 1956.

bubbles by many. It had graceful lines and presented a friendly face to the world. It was also very practical and, like other frugal cars, came into its own during the time of the Suez Crisis; it could stretch to 72mpg with ease. When fuel rationing came into effect in Britain, the country began to absorb the entire european delivery quota of Heinkels!

The Heinkel was produced in Germany, in left hand drive only, until 1958 when a production license was sold to Dundalk engineering in Northern Ireland. Dundalk added a convertible version to its production which was, again, totally left hand drive. There was also some smaller scale manufacture in Argentina.

In 1962, the Trojan Company of Croydon, England, began to make the car under their own name as the "Trojan 200". In all over 6000 Trojans were made, in both left and right hand drive. There was a van version of the Trojan 200, although only a handful was made: all right hand drive. The van had a steel body with a glassfibre top. Some Trojan models were made for export and these had four wheels: the rear pair set very close together.

Overleaf: A convertible version of the Heinkel was built in Northern Ireland; this example is seen with a Trojan of the same hue.

It may not be luxurious, but the interior is cheerful, well-equipped and amazingly spacious in relation to the overall dimensions of this tiny car.

Inside the convertible there's plenty of room for two adults at the front. With the hood up, the car retains the charm and appeal of its fixed roof siblings. A car for all seasons.

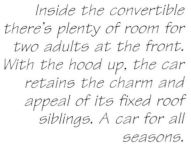

Previous pages: Because the steering column is not attached to the door - like that of the Isetta - it was not essential to change the side on which the door hinged when right hand drive was introduced. Once in, there is a surprising amount of space!

A major cosmetic difference between the Isetta and Heinkel/Trojan families is that the latter's headlamps are inset. This gives the Heinkel/Trojan an appealingly open 'face' and you can easily see why fans consider it the prettiest of all the bubbles.

Previous pages: The little convertible - German ancestors, built in Northern Ireland and pictured in England.

No wonder the description "Bubblecar" has stuck!

JANUS

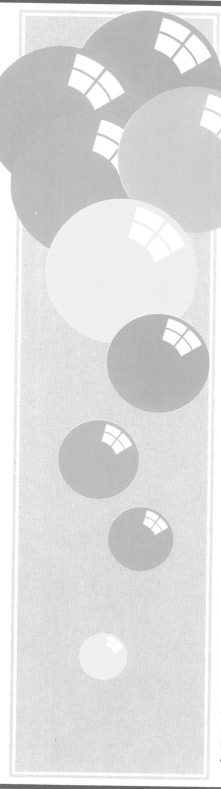

The Janus, at first glance, might well be back to front. Or even front to back, for it is very nearly symmetrical with a door at each end. Externally, the only immediately obvious differences between the front and rear of the Janus are different lamp housings for front and tail lights and deeper wheelarches at the front.

The car seats four on two bench seats. The front seat faces forward and the rear seat faces backwards, with doors at each end of the car for access. Netting in the door panels and space between the seats provide accommodation for small items of luggage: the rear seat can be removed to give extra space for larger loads. It is even possible to convert the seats into sleeping accommodation for two and still leave space at each end for luggage. Presumably, when it comes to sleeping arrangements, it pays not to be too tall ... The doors open at an alarming angle, thanks to the shape of the body ends; grab handles at both sides help with getting in and out of the Janus.

The engine, a 250cc Zundapp single-cylinder two-stroke, is placed right in the middle of the car between the axles. The gearbox has four forward and a reverse gear and is of a novel design, pioneered by Zundapp and the Getrag Gear Company, whereby all the gears - except reverse - are constantly in mesh. A suitably-shaped rod moves in the hollow gearshaft and forces steel balls from recesses in the shaft to lock the appropriate gear so that it rotates with the shaft. There is no chassis so everything is supported by the monocoque body structure.

The car takes its name from Janus the Roman god with two faces, each looking in different directions. He is the god of doors, which always have two sides, and gave his name to January which looks back to the old year and forward to the new. The Janus, designed by Claudius Dornier, was produced at the Zundapp works in Nurnberg. Although derived from the Dornier Delta the Janus is more rounded and bubble-like in shape than its precursor.

An interesting story about the Janus - at least for English

Less of a bubble, more of a car, the Janus is nevertheless small, with an overall length of just 2.86m. Like the Isetta and Heinkel/Trojan the Janus has a large door at the front - but also another at the back. The rear seat can be removed completely for larger loads. An early press photograph showed the Janus being loaded with a full-size refrigerator!

speakers - is that the "Janus" badge is in a fine chromed script: the capital "J" being joined to the body by a particularly thin piece of this metal script. Corrosion, or over-enthusiastic polishing, can cause the "J" to become detached and fall off, leaving a rather unfortunate new name ... For the last six months of production, the name was painted on instead.

The little Janus is quite lively to drive, with good springing and roadholding. Handling, too, is good, as might be expected given the equal distribution of weight which results from the central positioning of the engine. Maximum speed attainable is 53mph and the abstemious fuel consumption 65mpg.

Sadly, the Janus was produced for only a little over a year, during 1956/57. However, when the Zundapp factory was closed down thirty years later, tucked away at the back of the workshop was discovered a brand new unsold Janus that had never left the factory. "As new", this extraordinary survivor now graces a museum in Berlin.

When the bodywork is all of one colour, the affect of total symmetry is heightened, especially from a distance. The wheelarches give the game away - the baby can't drive.

Basic instruments - with one modern addition that could well save the day.

At the fulcrum - access to the amidships engine is under a hatch between fore and aft seats.

ISETTA

Unlike many other bubblecars, the story of the Isetta begins not with aeroplanes, but with refrigerators. Milan-based manufacturer Rennzo Rivolta decided it was time for a change of business direction and, in 1953, began making little cars. He called the model *Isetta* or, in other words, little Iso. It was this car that started the trend for bubbles. Indeed, the basic shape of the Isetta (along with that of the Heinkel), is what most people imagine when they picture a 'Bubble Car.'

In 1955 Rivolta sold the rights of manufacture of his little car to BMW who were, at the time, in a sorry financial state and were looking for a small, affordable car design to boost sales. BMW replaced Iso's original two-stroke engine with a modified version of one of their own motorbike engines, an air-cooled, 247cc single-cylinder four-stroke. It sat at the back of the car and drove the rear wheels via a gearbox and enclosed chain. BMW also redesigned the car's suspension. The body was mounted on a tubular chassis. There were two rear wheels set just 48cm apart. Isettas of this type were exported by BMW and those that found their way to England would have been imported and distributed by AFN.

In 1956, a 295cc engine was introduced for export models, giving the Isetta its "300" name. There were changes to the body style too, most notably the introduction of differently shaped sliding windows. A dual colour scheme also became available, with a chrome strip at the dividing point.

The front-opening door was a novelty when the car was launched, but also very practical. The steering column and instrument binnacle swings forward with the door, so access to the interior is easy. The Isetta seats two adults, or one adult plus two children, in comfort and has space for some luggage too. It's possible to park front end on to the kerb, and step straight out onto the pavement. The *Motor* road tester also pointed out that the door arrangement, combined with the car's small dimensions, would enable it to be housed in many a garden shed.

Buying the rights to make the 'Little Iso' marked the turnaround in BMW's postwar fortunes. This "300" from the Netherlands has the original dual colour scheme and sliding windows which were introduced in 1956 for export models.

The little car proved to be very popular and turned around BMW's postwar fortunes. Although for its time a daring and unusually designed car, the Isetta was generally well received and thought to have none of those annoyingly pointless additions and fiddles that plagued other small cars. The accessories were of high quality, the headlamps better than on many a larger and more expensive vehicle. The windscreen wiper earned praise for actually wiping all of its screen.

In 1957, a four-passenger version of the BMW Isetta was introduced. Access to the rear seats was via a right rear side door and a 585cc engine gave this version its "600" name.

Apart from BMW's production, Isettas were also made under license in France, Brazil and England. The British-built cars were promoted by Mr R J Ashley, an ex-pilot who had turned to motor manufacture. The cars were built at the old Brighton Locomotive works which, in March 1957, had ended production of steam locomotives. Two railway tracks ran through the works, making it ideal for conversion to small car manufacture: all parts had to be delivered by rail anyhow, as the works were situated at the top of a very steep hill and were unreachable by road. Within three weeks of work starting on the factory conversion, the production line for Isettas was ready to roll. Locomotive building staff were kept on - building rather smaller means of transport than they had been used to. Parts arrived from BMW in Germany and were assembled on a single production line, with a 'siding' into and out of the paint shop. At the end of the line, 25 new Isettas at a time would await train transport out to dealerships. Customers who preferred to do so could collect their car from their local railway station (pre-Beeching!), with almost zero delivery mileage.

The main change to the Isetta for the British market was the introduction of a three-wheeled, right hand drive version. This was particularly attractive, as road tax favoured the three wheeler considerably. The rhd model's door was hinged from the right hand side of the car and the steering column moved across to the right as well. Unfortunately, the rhd conversion put engine and driver, and hence most of the weight, on the same side of the car and to compensate a lead counterweight was placed behind the trim panel on the nearside. Lucas electrics were used in place of the German Hella alternative, with a different headlamp housing being used too. Advertising claimed that the Isetta was "... the world's cheapest car to buy and run and ... is the easiest car in the world to park."

When the Brighton factory began production it already had an advance order for 1000 4-wheeled cars for export to Canada. The car proved very popular there, as in other British Commonwealth countries. However, in 1958, the Canadian importers went bankrupt and unsold cars were returned to Brighton where they were converted into three-wheelers for the British market.

In addition to the basic Isetta car, the Brighton works also produced a pick-up truck with fabric roof, based on the three-wheeler, and a small four-wheeled delivery van, which was very popular on the continent. As a *Motor* road test pointed out "It is ... willing to go anywhere and do anything, and the manner of its going entertains some people very much indeed."

There were two particular Isettas which were very special cars indeed. They were used to smuggle refugees across the border from East to West Berlin when the city was divided. Because the Isetta was so small it was not searched by guards at the checkpoints: they assumed that no-one could ever hide in it. The two four-wheeled Isettas were converted to carry a hidden passenger alongside the engine. The air induction and heating systems were removed, the exhaust pipe re-routed, the rear mudguards removed and a metal frame lined with plywood was installed to hide the refugee. One of these Isettas was involved in six escapes, the other managed three but, on its fourth trip, the car was held up at the border and the guards spotted movement of the car as they checked the driver's papers. The car was taken apart and both refugee and driver arrested. The remaining Isetta is now on display at the "Checkpoint Charlie Museum" in Berlin.

The typographical style of door badges changes dramatically ...

Two Brighton-built Isettas, three wheels for the home market, four wheels for export, both right hand drive.

.. but the BMW logo only very slightly, from gold to silver.

Overleaf: Many British-built cars were fitted with these chromed tubular vertical bumpers or "Nerf Bars". The only problem was that the top part of each bar was simply bolted through the bodywork so a collision invariably meant that the wing got dented by the bumper bar itself!

Right & left: Detail of right- and left-hand-drive Isettas. Note the single semi-circular direction indicator to be found on each side of the cabin.

Red bubble or blue bubble? Decisions, decisions!

It's easy to fall in love with the cute Isetta. This particular example has the luxury of twin windscreen wipers!

Kermit Koupe?

Owning a classic car of any type is no picnic. Because original spares are hard to come by, owners tend to buy parts as and when they are available rather than waiting until they are needed. This picnic basket actually contains an extensive tool kit and essential spares for running repairs.

The Isetta's fixtures and fittings are well-designed, of very high quality, and last well.

This lovingly restored Isetta features period accessories including chrome headlamp hoods, luggage rack and mudflaps. Whitewall tyres add further style.

When the door is opened, the Isetta's steering wheel and shaft, attached by a universal joint, moves upwards and outwards with it, making getting in and out trouble-free. There is a surprising amount of legroom for driver and passenger. The gearchange is difficult to master initially, requiring a light touch.

BMW had the ideal power unit for their new acquisition in the shape of their proven single-cylinder motorcycle engine.

LADYBIRD 4

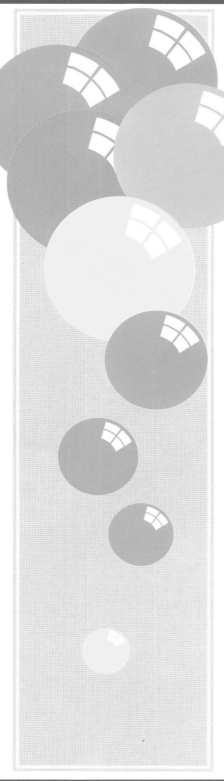

In 1960, Edward Turner, Managing Director of the BSA Group's Automotive Division, began to look at the possibility of a small car that would take on the German bubblecars and rival British small cars. He already had the ideal engine in the 250cc BSA Sunbeam/Triumph Tigress Scooter unit (an engine which had also provided power for a front wheel drive milkfloat). The plan was to use the same transmission and rear suspension as the scooter's too.

The first prototype consisted of a chassis and frame of steel tubing covered by a hand-beaten steel body: the latter was made by Ben Johnson of Carbodies Ltd which was part of BSA. Constructed in just nine weeks, the car was ready by the end of September 1960. Its shape is beautifully sweet; amazing to consider that those curves were created entirely by hand. Unfortunately, such complex shapes would have been too expensive to press for a mass-production vehicle, so Carbodies were asked to produce a second prototype with panels that could be mass-produced more easily.

This second version, which had a hinged hard top, was finished two years after its predecessor. But bubblecars were going out of fashion by then and Alec Issogonis's Mini delivered everything that bubbles had offered, cheaply, and in 'real car' form. BSA's scooters were being phased out anyhow, and Mr Turner's retirement approached. This chapter of motoring history was drawing to a close.

The two prototypes were left to collect cobwebs. Eventually number one went to the USA and, after restoration, found itself back in the land of its birth in late 1991: it now forms part of the Mike Webster collection. Prototype number two moved to Folkstone and disappeared from view. For the time being at least, its whereabouts are uncertain.

The number one car, which is our photographic subject, has independent suspension and handlebars for steering. The hatch on the rear deck gives access to engine and fuel tank. The engine has an electric starter with pull switch on the cabin floor; there is a hand operated starter lever for emergency

Beautiful to look at but too expensive to mass-produce - the ladybird's curves are a work of art.

Often seen on two-wheeled transport, the emblem of BSA Cycles Ltd, which the ladybird wears above the rear light.

The Ladybird wears this BSA badge on its bonnet.

What a shame they were never produced in quantity; just think - swarms of Ladybirds!

use. The brakes work on all three wheels and are compensated to ensure even braking at each wheel. The car does not have indicators or a windscreen wiper, so it has probably not been used on public roads a great deal although it is registered for road use. Unlike prototype number two, the earlier car never had any protection against the elements for the occupants.

    The car's development was shrouded in secrecy. At the time very few people even knew of the project's existence, let alone saw the car. The "Ladybird" name was not an official title but so appropriate was the car's nickname - imagination easily adds the black spots - that it stuck.

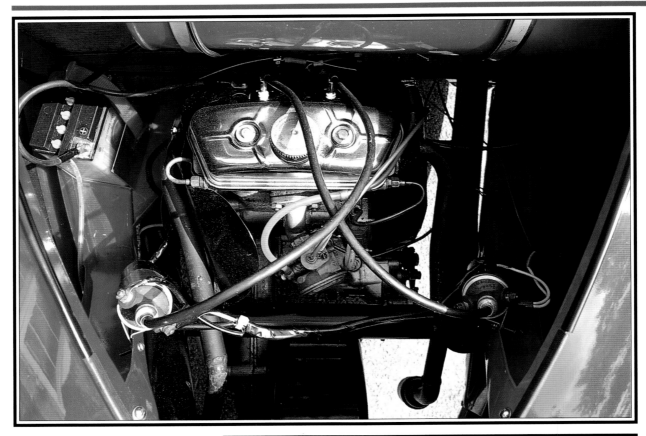

The Ladybird was designed to find new uses for BSA's 250cc scooter engine, which sits behind the car's occupants and ahead of the rear wheel.

The car's switchgear is borrowed straight from the BSA Sunbeam Scooter. The pull start switch is on the floor in front of the driver. The central controls are the emergency starting lever, gearlever and handbrake.

Previous pages: The cost of producing the prototype must have been enormous, but records show that, in 1962, the production car would have sold for £283.

MESSERSCHMITT 5

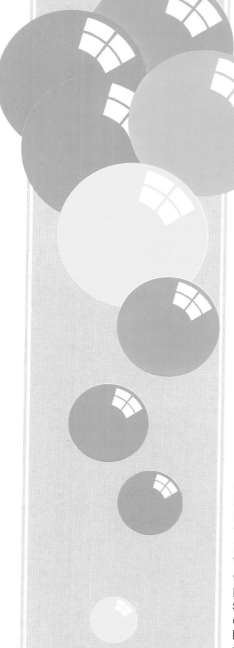

There is no truth in the story that Messerschmitts were designed to use up a surplus of aircraft parts after the end of the war. Nothing would have pleased Fritz Fend more if it had been true. Instead, he had to make do and mend, and was thwarted at almost every stage of his vehicle-building plans by a chronic lack of everything he needed.

During the war, Fend worked at the Messerschmitt aircraft factory involved in the development of Germany's first jet aircraft. The idea of creating a new form of transport had come to Fend shortly after he had moved to Rosenheim to take over the running of his father's shop. At first he worked on a covered, pedal-powered tricycle for which he designed a unique gearing system. As he had no particular knowledge of the subject and had no data from any other source to work with he devised, and performed, a series of one-man tests. He chose a particular flight of stairs to run up carrying various weights and ascending one step at a time (low gear) or two or three at a time (high gear). By this method he managed to collect enough data to judge how his tricycle would perform going up hills. He then proceeded to produce a working model, which proved the whole tricycle idea sound but, unfortunately, materials and machine tools were extremely difficult to get hold of at the time. Fend continued to mull over his project at the back of his mind, as the need to live forced him to concentrate on the family business.

In 1947, Fend had a visit from a war veteran who had lost both his legs and was only able to move about with the aid of a small wheeled trolley. If a suitable means of transport could be built, Fend reasoned, and the War Wounded Association were interested, materials might be forthcoming. Producing a vehicle was hard work, but he made it. The resulting small three-wheeled tricycle had a seat in the middle and a control lever with handlebars in front. Pushing and pulling on the control lever powered the car; steering was via handlebars. The interest he had hoped for materialized. The Association ordered fifty of his

Left: The plexiglass canopy of the KR175 earned it nicknames - a sure sign that it had been taken to people's hearts - such as "Snow White's Coffin" and "The Cheese Cover." This car features the original flat windscreen.

pedal-powered "Disabled Driver's Vehicles", although the amount the organisation paid him produced very little profit and did nothing to build up this side of his business. Fend persevered and went on to produce a version of his tricycle with a covered body. A first rate selling job on the Association, and the interest of the Ministry of Labour in Munich, gave Fend the opportunity, at last, to start producing his vehicle with proper materials.

In 1948, Fend bought a power-assisted bicycle with the aim of using its 35cc engine in his "Flitzer", as the tricycle had become known. The first test drive produced an amazing 40kph! Suspension and steering needed a rethink though. His next test vehicle was fitted with a 100cc engine made by Fichtel and Sachs, who were very impressed by the machine and consequently struck up an excellent and enduring business relationship with Fend. The bicycle-type wheels of the prototype were not really able to cope with the performance Fend was now achieving, but car wheels were too large: his first solution was to use wheelbarrow wheels at the front and the original bicycle wheel

Below the fuel tank is the Fichtel and Sachs 175cc engine. Support from this company had been very important to Fend in the early days.

Overleaf: The aeroplane-style of tandem seating is not to everyone's liking, but it uses space efficiently. Inset shows KR175 cockpit with handlebar steering: the leather strap prevents the canopy from opening too far.

Left: The last of the KR175s were fitted with the new, rounded windscreen, style of canopy which was employed on the KR200 model.

'Suspension' on the KR175 meant a large spring under the driver's seat ...

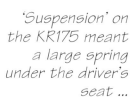

Following the trouble with Mercedes, the three-circle logo replaced the stylised bird to which they objected. The firm of Krupps then objected to the three-circle logo, so it was changed again, this time for three diamonds.

A KR200 with the new style of curved winscreen canopy. The model incorporated many other improvements.

at the back. He continued to test further prototypes, some of which had very little in the way of equipment - no brakes for example. Stopping involved ripping the lead away from the spark plug so, presumably, planning ahead was important ...

Gradually Fend made real headway with his designs and was able to persuade Dunlop to make small tyres for his machines which were, by then, sporting three wheels of the same size. Fend was now producing ten cars a month, the maximum that he could handle without expansion. He looked for, and found, a partner to put money into the company but such an arrangement did not work out too well. The hardest work came from Fend, but the profits always seemed to reach the partner's pockets first. Business and personal disaster loomed. However, acting on a sound piece of advice, Fend contacted someone for whom he had worked during the war, someone who had been very impressed with his innovative engineering ideas - Professor Willi Messershmitt of Regensburg.

The postwar situation dictated that Messerschmitt was no longer able to produce aircraft so the company had diversified into other engineering projects and repairs. But Prof. Messerschmitt was ever on the lookout for something to fill all that empty space in his factory. Messerschmitt and Fend sat down together to discuss the car that they would build. It would be a three-wheeled,

The KR200 engine.

two-seater, with the passenger sitting behind the driver. A plexiglass canopy would offer weather protection and would be hinged at one side so that it could be lifted to give access to the cabin. The new car would be powered by a 150cc engine. The two men were convinced that they had a successful design, so a proper partnership was formed in 1952. By the summer of the same year a prototype was ready for testing. It was named the "Fend Kabinenroller" type FK150. The public were able to buy the car from 1953, when it was displayed at the Geneva motor show. A higher-powered version, the KR175 appeared soon after.

The new Fend/Messerschmitt car was very popular, both with the press and the public. The see-through canopy was a very novel concept, an obvious reminder of aeronautical days for some. There were criticisms - of high noise level and vibration in particular. 1955 brought a power increase from an engine of 200cc, the KR200. The major problems had been cured now, and body changes gave a much smarter appearance. The car's front track was increased for greater stability and new interior options promised better comfort. At this point the canopy was modified to provide a new, more modern, curved windscreen. The car got a proper suspension at this stage too, doing away with the springs under the seats which had previously provided some crude suspension for the occupants. The upgrading had been successful; a motoring journalist at the time described the car as "The jet for the everyday man ... all in all the KR200 is the only car on the market at this price which actually achieves a perfect harmony between engine and chassis". The car continued to be just as popular as ever, even as more competition from other small cars affected the market.

Fend decided to show the world what his little car was really made of. He prepared a KR200 to attempt speed records and, on August 29th, 1955, it set off around the Hockenheim circuit to show its true colours over a twenty-four hour test. The lap speed average settled down at around 106kph. And there it stayed - lap after lap - hour after hour - for the entire twenty-four hours. The record that the car had broken had been set seven years previ-

Above and overleaf: A rare KR200 convertible. The canopy lifts in the normal way for access. Overleaf inset: detail of the front indicator light.

The interior of this car is particularly fine - and fun too. White piping and stitching on red seats: polished wooden side strips.

ously by a racing car with considerably more power. In all, Messerschmitt were to break more than twenty-five land speed records in their class.

Whatever the factory's success with the outstanding speed-cars, there were unfortunately problems with the standard product. Build quality had deteriorated and customers were dissatisfied. As if this were not enough, Mercedes took Fend to court, claiming that the logo he used

The KR200 steering bar looks like an aeroplane's controls. Note the original radio and how the Mercedes-like original logo continued to be used inside the car.

The canopy has sliding side windows but, in the sunshine, especially when stationary, the greenhouse effect is considerable.

was too similar to their own. It is difficult to see the confusion - and one wonders what difference it made anyway - but the larger company won the day. A new logo with the letters "FMR" inset in three rings was introduced.

By now Messerschmitt's car arm was in serious debt. The German Government were looking towards starting production of aircraft again and Messerschmitt would be an ideal choice. The debt was written off, but demand for all *Kleinwagen* was falling anyway and the company's production problems didn't help - so money continued to be lost. Professor Messerschmitt was given an ultimatum. Government money could not be allowed to shore up the ailing car department when it was intended for aircraft development. The car operation was grossly unprofitable, so it had to go. The car arm was bought by the state of Bavaria and split into smaller manageable units. Messerschmitt kindly allowed Fend to continue using his company's name and logo even after the split.

When the Regensburg factory came up for sale, there was much interest, and eventually a deal was made in which Fend and businessman Valentine Knott joined forces to buy it. It was a shame for Fend that such a situation had not come about earlier. The day of the larger car had arrived. Still, Fend was an enthusiast as well as a businessman: he looked for cheaper methods of production, of the Plexiglass dome, for example, and made every effort to improve build quality and keep production going at an economic level. He also had plans to build other vehicles - a twin engined jeep-type all-terrain vehicle was on the drawing board, but it got no further than that. He also worked on a new sports version of the Kabinroller to capitalise on the Hockenheim successes which had captured public attention.

When it came to creating a new sports model cost was, as usual, a deciding and very limiting factor. There was no chance of being able to design a completely new model for the general market. Instead, the existing three wheeler was modified by the substitution of two rear wheels and the track was widened slightly. The new four-wheeled car was unveiled at the Frankfurt Motor Show in 1957. There were two versions, with 400cc

The rare convertible model is, today, much sought after by enthusiasts.

Previous page: Everything in the garden's lovely.

This is a deluxe version with smart two-tone paintwork.

and 500cc engines. With this car (known as the "Tiger", despite further copyright wrangles), a decade of three-wheeled Fend/Messerschmitt motoring came to an end.

With the convertible's hood up, the only real drawback is impaired rear vision.

Overleaf: Many makes of bubblecar owed their existence either directly, or indirectly, to the influence of the aircraft industry - the Messerschmitt was no exception.

PEEL TRIDENT 6

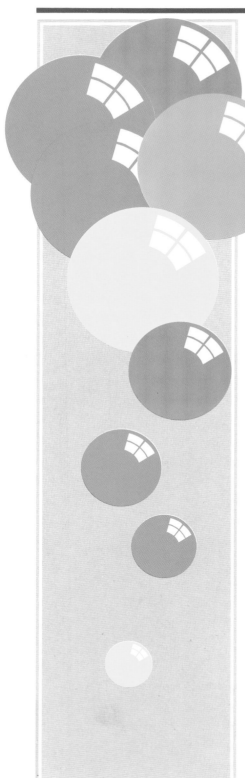

It is perhaps appropriate that a very small country with a three-legged emblem should produce a very small car with three wheels. Peel Engineering Ltd. began small car production in 1962 with the P50: a tiny, upright, single-seater car. By 1965 the car had been completely redesigned as the Peel Trident, a car with a slightly larger and more bubble-like shape than the P50. There were single-seater and two-seater options.

The body of the Trident is in two sections. All the mechanical components are attached to the lower half. The upper part, topped by a dome made of Plexiglass, is hinged to the lower half of the body and is raised to give access to the cabin.

The engine used in the Peel Trident is a DKW 49cc two-stroke which drives the rear wheel. The Trident's tiny wheels are just five inches in diameter!

Single- and two-seater versions look the same because the basic design is almost identical. However, in reality, the single-seater is slightly smaller: the second seat is replaced by a container for stowing parcels or small items of luggage. Both versions of the Trident also have stowage space at the rear.

Peel made an export model of the Trident with a 125cc engine, centrifugal clutch and larger rear wheel, although production numbered but a few. Indeed, the total production of Tridents was no more than seventy-five cars.

Opposite: In a remarkably small machine there is space to carry luggage at the rear, and room for a passenger in front.

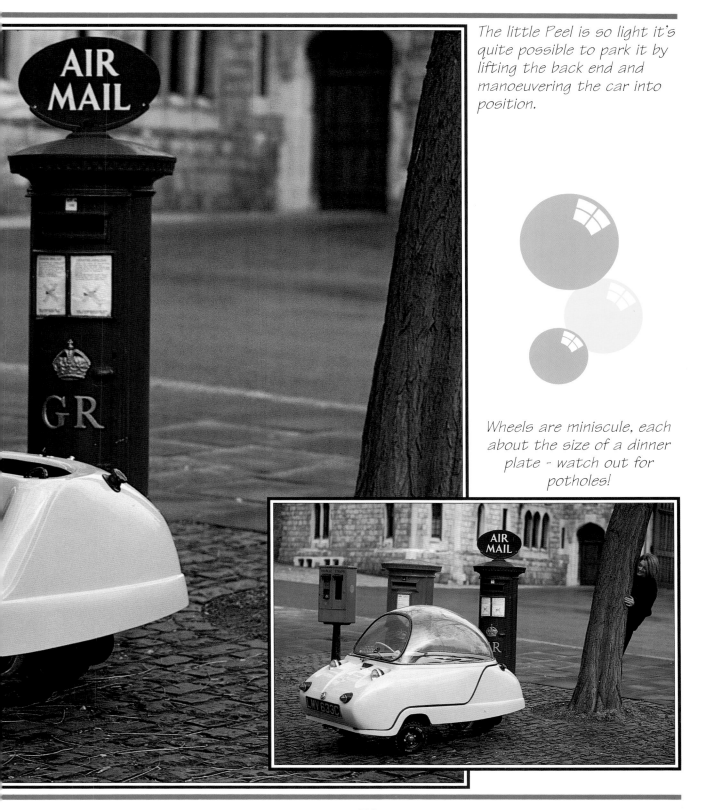

The little Peel is so light it's quite possible to park it by lifting the back end and manoeuvering the car into position.

Wheels are miniscule, each about the size of a dinner plate - watch out for potholes!

Although the P50 is a square box, the Trident has looks that represent the jet-age! While the car was being photographed in Windsor, it was surrounded by Japanese tourists who, having never seen a Trident before, assumed it to be a new car and were absolutely intrigued by it.

Left: One of the problems with the Trident is misting-up as there is almost no airflow inside the cabin.

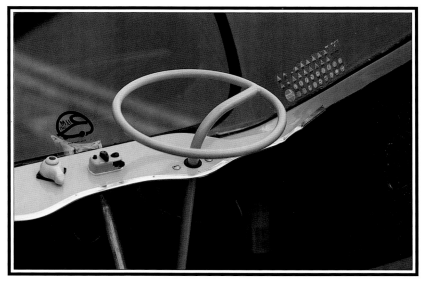

The most basic of controls including a decidedly Citroenesque steering wheel.

SHELTER 7

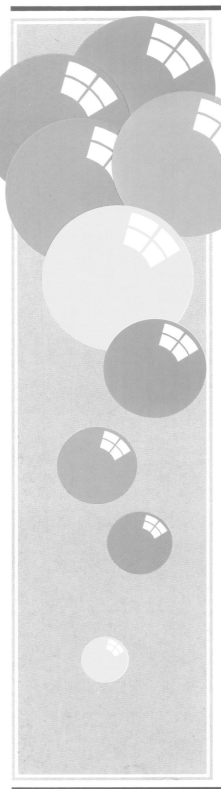

If ever the epithet "handmade" suited a car down to the ground, it is the Shelter. This little car began life as the university engineering project of Arnold van der Groot, who had worked for the Bristol Aircraft Company after the war. His idea was to produce a city car that could be used in the conurbations of Amsterdam and elsewhere. Cars would be hired, driven to the destination drop-off point and left there for the next user.

The project was begun in 1954 and, after a lengthy period of design and development, was ready in 1956. The Netherlands government expressed an interest in the cars; it hoped to sponsor the project.

Van der Groot designed, engineered and constructed the majority of the parts for the car himself, including the engine which was a 228cc single-cylinder two-stroke. It gave 8bhp at 4200rpm. He built the car's suspension from scratch. A few parts, the speedometer, for example, were bought in, but they were decidedly in the minority. The steering wheel was handmade and body panels were mainly flat sheet steel, curves being made with a roller where necessary. Indeed, so unsculptured was the *carrosserie* that it was decided to introduce a rounded look into the headlamp surrounds: such a shape was a real problem, but a trip to the local saucepan factory soon cooked up a suitably-shaped solution.

Van der Groot bought the coils for the dyna-start starter (combined starter and dynamo), but constructed his own housing for cheapness. The engine's connecting rods were constructed from pieces of gas pipe shaped on a simple bending machine and welded together. Such constructions were spot welded, Mr Van der Groot being the proud owner of large and solidly made vintage 1950s

The Netherlands government was keen on the idea of a city/commuter car - an idea that has been tried in the Netherlands and other countries since, although with limited success.

The minimum of interior equipment and even the steering wheel was handmade.

With the rear panel removed access to the engine and spare wheel is easy. In fact, an engine change takes only five minutes: a real boon if the 'commuter-car' idea had worked out.

This Shelter is in the process of being restored by Sjoerd Ter Burg in Winterswijk, Holland. It shows the rear panelwork in place, and those 11cm saucepans (minus handles) which form the headlamp recesses.

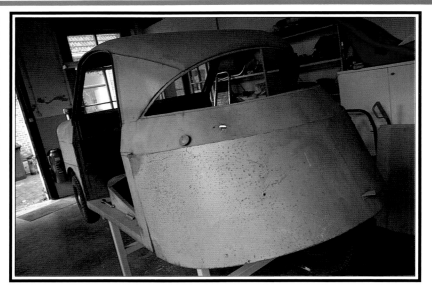

welding equipment reminiscent in style of American cars from the same period.

The car's curved roof was formed by an ingenious water press. A concrete mould was made, and the sheet steel for the roof was placed over it, followed by a thick metal plate which was clamped in place. Water was pumped between the heavy plate and the thinner roof panel forcing the latter into the mould.

The positive attributes of the car were to be cheapness, lightness and ease of maintenance - it took only five minutes to remove the engine and replace it with another. Initial problems with the connecting rods were solved; unfortunately though, there were other, quite serious drawbacks. Due to an engineering oversight, after 200 miles or so, the rear axle would simply sheer off: the geometric relationships between back wheel and engine were wrong. Bumps in the road would create such twisting torque that the axle would give way. This fault would have taken a complete redesign to cure, which was simply not economically viable. The Shelter's other major fault was even more worrying ... the car had an alarming and dangerous tendency to suddenly burst into flames!

As a consequence of these development problems, the Shelter project was never really going to achieve full scale production. Although by 1958 sufficient parts were available to make twenty cars, only seven were ever finished: all made at the designer's home.

The Shelter is 2.20m in length, 1.06m wide and 1.16m high. It weighs just 220 kilos.

VELOREX

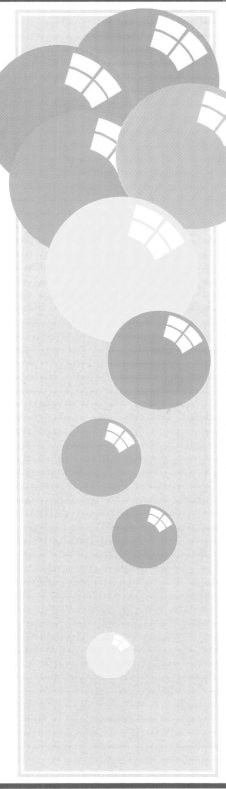

At first sight the Velorex suggests the result of a union between a frame tent and a flying helmet. Hailing from the Czechoslovakian town of Hradec Kralove, the car, which was produced between 1953 and 1971, was fitted with a rear-mounted Jawa two-stroke air-cooled motorcycle engine. Three engine sizes were used. The 125cc and 250cc versions, whose engines were both single cylinder, had the motorcycle kick-starter converted for hand operation. The third type, the 250cc twin cylinder, had, in addition, an electrical starter and 12 volt electrics. Several components came courtesy of Skoda. The Velorex's terrifying top speed is 130Km/h (81mph)!

Built on a frame of welded steel tubing, the Velorex's 'bodywork' consists of wooden panels and leathercloth. Mudguards are metal. The car has four forward gears but, since the engine is able to run backwards, no conventional reverse gear is necessary and, in fact, four reverse speeds are available too!

In 1959, a four-wheeled car was produced by the same team. It was not an adaptation of the Velorex, but a completely different design. Styling was more modern, a 350cc engine was fitted and many Skoda parts were used. However, production costs were high and so the car never got into production. In addition, one three-wheeled pick-up version of the Velorex was built for use as internal transport within the factory.

You have to be eccentric, in the nicest possible way, to be a bubblecar fanatic, whatever the make. However, ownership of a Velorex takes other qualities too - courage and a thick skin for starters. Dick Boerstoel reports -

"Driving a Velorex is an odd experience. It starts with the way it looks; so different from any other car. To have less car would be impossible. When one is tall, taller than average, it is not easy to get into the car. The big steering wheel and the short distance to the pedals prevents quick hopping in and out. But as soon as one is sitting in the car, it is like entering a different world. The design is straightforward with no luxury -

except the ashtray that is!

"After a few turns of the dynastart the engine starts with an enormous uproar. Into first gear, easy on the clutch, step on the gas, and away you go. The acceleration is fantastic. Up to 60Km/h (37mph) a Velorex is far and away faster than a modern car. Above that, the acceleration is less impressive but, by then, all other road users are impressed. That is one of the nice things about Velorex-driving. At first people take pity on one, looking helpless between big cars and lorries but, after your hell-of-a-start, they are speechless ...

"A Velorex asks for hard driving. Fast turns are a joy - but make them only with a passenger on board. I once did it too fast on my own. My neighbour saw me, and said afterwards: 'You were driving on two wheels - I could see the bottom!'. I did not tell him that it scared the living daylights out of me.

"It's a pity that the Velorex is not fully reliable; the electrical system especially is weak. The battery often goes flat, a problem which requires some improvisation when it happens in the middle of nowhere. You have to ask people politely for a push, or a tow. I've got to know all the escape lanes, laybys, car parks and roadside emergency 'phones in the Netherlands. Once, it even

Dick Boerstoel's car before the unexpected upside-down meeting with a Dutch road.

Definitely not a car for the shy and retiring: the Velorex attracts attention wherever it goes. Just like any microcar - only more so!

Previous pages: The frame of the velorex is made of welded steel tubing. the external covering is leathercloth.

With the cover removed to expose the framework, the nose looks like that of an aircraft.

The fabric roof and body ensure that the Velorex has excellent power to weight characteristics.

happened that the spark plugs melted! In Czechoslovakia, the road conditions rule out driving at high speeds and for long distances; so perhaps this kind of use is the cause of the frequent troubles?

"Not all troubles are caused by the Velorex. I had a nasty experience on a narrow, winding road when passing an oncoming car forced us both to drive slightly off the road. After we had passed each other I tried to steer back onto the road, but the edge of the road was too high. At 40Km/h (25mph) the nearside front wheel was 'tramlining' along the edge of the road and this made the car lurch which, in turn, made the single back wheel slip off the road. At that very moment, the front wheel regained some grip and climbed back onto the road - this made the Velorex take off like an aeroplane and it spun around like a top with only the rear end touching the ground. After one-and-a-half turns it landed upside down: the Velorex was severely damaged.

"To Velorex, or not to

The design of the interior may be straightforward and luxury-free, but it does have a certain style!

Velorex, that is the question. The English language is herewith expanded with the following verb 'to velorex'. I velorex, I will velorex, I velorexed. In my case, the present tense usage is now very rare, the past more common."

Left & overleaf: With its fabric-covered body and narrow, spoked wheels, the Velorex harks back to the vintage cyclecar period.

Dear Reader,
We hope you enjoyed
this Veloce Publishing
production.
If you have ideas for
other books on automotive subjects, please write
and tell us.
Meantime, Happy
Motoring!

Postscript
Many of the people I have met while working on this, and other, books show a great interest in the way I photograph their cars. Inevitably the conversation turns to cameras, and all the paraphernalia associated with making photographs. I suppose it is also an interest in things mechanical, and very often people like to look through my camera at their car! For seventeen years I have worked with Leica cameras, changing them as different options become available. For me, the choice of Leica lenses is paramount. I depend on them. My favourite film is Kodachrome. I hope you enjoy the results.

David Sparrow